Scientific Writing for Students and Young Scientists

Dr. Alfred Orina Isaac

ISBN: 151428961X
ISBN-13: 9781514289617

To:

Mang'uti, Martha, Barack, Jacelyn and Jorey

CONTENTS

ACKNOWLEDGEMENT

I acknowledge Prof J.C.K. Lai (Idaho State University) for his dedication in teaching the art of scientific writing. I benefited tremendously from his knowledge. I acknowledge Mr. Isaac Mang'uti Keengu for his love for education.

PREFACE

It has always been a desire to write a guide for scientific writing that is brief and informative, to aid students and scientists. The aim was to make the guide adequately brief such that someone for example, on a three hour flight to a science conference, can read it in its entirety. This guide fulfils that dream and will benefit students or young scientists in most science oriented disciplines such as Biological and Biomedical Sciences, Medicine and Agriculture. It also provides an incredible resource for those teaching research methods at various levels.

This guide provides a simple but detailed step by step breakdown on how to write the various sections commonly found in scientific papers namely: the title, abstract, introduction, methods, results and discussion. It will help students and researchers at the start of their career approach their writing tasks with greater confidence and therefore realize greater success as scientists. In addition to helping science graduate students and young scientists, this guide has been simplified to be resourceful to undergraduate students, who are expected to write a science research project.

Over the years as a graduate student, lecturer and scientist, I have recognized a remarkable lack of basic scientific writing skills by students and even scientists. This gap is evident not only with undergraduate students, but also with those pursuing postgraduate education. Moreover, depending on the rigor of supervision, some individuals with doctoral degree training exhibit poor knowledge of the structure of various components of most scientific documents. Due to these deficits, most graduate students have their research proposals rejected, derailing their ambitions for higher education. Furthermore, such rejection also manifests in scientific manuscript submissions and grant applications. It is important to note that some of the various sections of scientific papers and reviews are similar to those found in thesis/dissertations or grant applications.

This guide's format and structure is presented in a more general way. It is possible that various Journals and disciplines may utilize slightly different formats or styles. Nevertheless, a good grasp of this format and structure will enable you to easily relate and adapt to any other style you might encounter in your writing.

However, I would like to point out that no single guide will solve all writing challenges and make somebody an excellent communicator, unless that person puts what is learnt into practice. As stated by Matthews (2008), *"learning to write skillfully is, always has been, and must continue to be hands-on experience"*. Practice makes perfect. In this spirit, mentors and laboratory heads are encouraged to always give their graduate students and post-doctoral fellows an opportunity to write manuscripts under their keen guidance. Otherwise, how else will they learn to write well?

NOTE: The organizational structure and various strategies presented here for writing scientific papers and thesis/dissertation, can be applied in writing book chapters and reports.

Chapter 1

IMPORTANT CONSIDERATIONS IN SCIENTIFIC WRITING

I t is considered that before you start writing, you must establish that your research meets certain minimum criteria. This must include results that are consistent, reproducible and with reasonable contribution to existing knowledge.

A scientific paper must demonstrate thoughtful organization, accuracy and logic. The author must be convincing and critically aware of the audience at all times. **To achieve this, embrace the following important general considerations:**

The first impression
Will you capture the attention of a very busy Professor reviewing your paper? Write the title, abstract and introduction sections critically aware of this phenomenon. You must write to capture the first impression.

The storyline/theme
Always inform the readers the purpose of your research and your central message. They should clearly understand or take note of the major breakthroughs in your work, if any. This ensures that you capture the reader's attention and interest and their imaginations.

Flow of ideas

As you write your paper or thesis, ensure that ideas flow logically from one section to another. This consistency must be maintained throughout the document. For example, the findings from your first objective must be presented as the first item in your results section, and then the rest to follow in the order they are listed. Similarly, information in the introduction and discussion should maintain that flow of information and ideas. Note that even the abstract, which is the summary of the entire document, is written with this order in mind. This ensures that your readers are not confused at any point, and they can link or relate information presented as they move from one section to the other. Upsetting this balance usually results in disinterest in your document and inability to decipher the storyline. A good scientist should always ensure that they mention their most significant findings first. This is encouraged and should not disturb the aforementioned sequence. Always remember that poor organization of ideas in your write-up can be interpreted to mean poor intellect and understanding and therefore poor research. To avoid this, you must write to the best of your ability in a clear and systematic manner.

Grammar and punctuation

Good grammar is like a glue that holds your entire paper together. A well written paper with proper grammar and punctuation, takes little effort to read and understand. When presenting a scientific writing, you should not burden your readers but strive to make their reading easier and enjoyable. It is of uttermost importance to ensure that your document is free from grammatical errors and poor use of punctuations. Presence of such errors especially in the preliminary and introduction sections of your documents will result in a negative outlook of the rest of the document regardless of whether subsequent sections are superbly presented. It is no surprise that top reviewers of major scientific journals and funding organizations reject documents out-right due to grammatical mistakes. Those mistakes are costly on any part of the document and reflects negatively on the author of that document. Once that negative image is created, everything else does not matter. This is because you have lost the reader, editor or examiner. Errors can be avoided if a document is shared with colleagues, diligent revision of various editions and use of spell-check available on most computers. It is also important for the author to be critically aware of the language requirements for their journal; that is the British versus the American versions of English.

Always strive to avoid long and winding sentences. Use short and concise statements.

To have a good document, avoid the following:

a) Do not use passive voice in all scientific writings.

b) Avoid use of informal language and colloquial expressions such as; "kind of", "the hell", "sort of" among others.

c) Avoid using first and second person style. Do not use "I", "we", "us"; which should be avoided and replaced with words such as "individuals", "participants", "research team". For example the use of the words "my findings suggest…" can be replaced with "findings in this study suggest…" Avoid the use of the word "you" to refer to the reader. Replace "you" with words such as "individual" or "person". Note that some schools of thought allow very limited use of first or second person in scientific writing. This only occurs when making references to unique events in the study only attributable to the authors(s).

d) Do not use exclamation marks and contractions such as "haven't" or "wasn't". Instead use "have not" and "was not" respectively.

Most writers like to use words such as "this", "it", "that", "they", "their" and "them" to avoid repetition in their statements. In formal scientific writing, clarity is more important and hence it is better to repeat certain words than create ambiguity by using the word "this". In brief, avoid using such words in formal scientific writing.

Do not use hyperbole in your writing under any circumstances. An example of a statement using hyperbole is: *"These findings are going to sweep the scientific community like a Tsunami".*

Most computers have facilities for checking spelling and grammar. Use them but cautiously because a human being understands grammar better than a computer. Rely more on personal judgement when correcting grammar. Always do a spell check for every paper or document before submission.

What is the role of punctuations in scientific writing? Punctuation helps create clarity and organization in your writings. Use punctuations to create emphasis and enhance useful suspense. Good writers use simple punctuations and avoid complex ones. For example, it is easier to separate sentences using a period instead of a semicolon. When used appropriately, commas create clarity and emphasis to a sentence. Pay particular attention to use of commas when using the words "that" and "which". A comma is put before the word "which", but is not put before "that". Rules to do with use of quotation marks and hyphens are changing and are quite complex. Whichever rules you decide to use, ensure consistency is maintained. If need be, consult a language specialist to ensure that your punctuations are correct. They are too important to be taken lightly.

Be brief

It is considered polite to be brief. If the paper is unnecessarily long, you owe your readers an apology! Peer reviewers delight in reading short and concise papers. In scientific writing, the quality of data or results is more important than their quantity. If you can communicate a specific scientific concept in ten words, do not use thirty. You should put into consideration that longer manuscripts, thesis or dissertations stay longer with peer reviewers because more time is required to read them; a scenario that can be unfavorable to the author. Factor in use of visual aids, which can help shorten your document significantly. A whole page of text can be replaced with an informative figure occupying a small part of a page.

Unless it is a requirement by your target Journal, do not justify (align) your text.

Use of abbreviations

Abbreviations form an important component of scientific writing due to existence of many complex and quite lengthy technical terms. Define all abbreviations the first time they are used. However, do not use an abbreviation unless a term is used at least three times; with the exception of the degree and percent symbols. To avoid the inconvenience of your readers flipping back and forth, expand letters the first time such a word is mentioned in every page. Note that some journals require that all abbreviations are listed at the beginning of the paper for easy accessibility by the reader. Unless extremely necessary, avoid use of abbreviations in the headings or sub-headings. Some of your audience might prefer to glance at those headings in order to make up their mind about reading the paper. An unfamiliar acronym or abbreviation in the heading is most likely going to displease such a person.

Abbreviations cannot be written in plural e.g. "five *mins*" or "four mls"; instead should be written correctly as "five *min*" or *four ml*. Use standard abbreviations for hour (hr.), minute (min) and second (sec); do not write the complete words. A list of some commonly used abbreviations that do not require a definition is attached (appendix 1). Note that currently, complete words are increasingly preferred to *Latin* or *Greek* abbreviations such as "i.e. (that is), e.g. (for example), and etc. (so forth)".

Numbers

Many Science Journals recommend use of Arabic numerals over Roman numerals whenever numbers are used. However, refer to your journal of choice for guidelines, and maintain consistence regardless of your choice of number style. It is true to say that numbers provide an invaluable tool in science. Just like words, numbers tell a story. How many scientific investigations are devoid of any numbers? Probably very few. Ensure that you do not start sentences with numbers. For percentages, it is recommended that the text includes the actual number for every percent figure stated. *For example; "20% (20/100) of the students smoked cigarettes".*

Use of parenthesis

Do not use double parenthesis. For example, *"catechins reverse the antioxidant effect of anthocyanins (Figure 2) (Table 3)".* This can be correctly written as; *"catechins reverse the antioxidant effect of anthocyanins as shown in figure 2 and table 3".*

Study and cite literature from distinguished scientists

Make a habit of reading articles from distinguished scientists in your discipline. In addition, it is good practice to access the most recent publications for the target journal. Use them as a stylistic and formatting guide in the writing process. You will be surprised how much you can learn from such articles and how they can positively influence your writing skills. Believe it or not, science has its own politics. Do not forget to cite your advisor's relevant publications, especially if you are a student!

Use of tense

Using the appropriate tense is very important in scientific writing because it indicates the status of the research being referred to. Present tense is used to refer to work that has already been published and is available to the scientific community. This as well applies to your own, previously published research. On the other hand, research that has not been published, as is the case with your data, is reported in the past tense. This distinction must be reflected throughout your paper. The exception to this rule is when you need to refer readers to specific figures/pictures or tables in your result section. For this case, use present tense. For example:

"Anthocyanins elevate glutathione concentrations as shown in figure 2".
"See figure 1, showing an illustration of antioxidant activity of anthocyanins".

Important results should stand out

Ensure that important findings stand out. This can be done by use of attractive and informative illustrations, or by making pointed reference to such results whenever possible without appearing too repetitive. Additional tactics to make certain aspects of your paper stand out is use of underline or italics, a numbered list and change of font. To help capture your readers' attention to particular results, use words such as: *dramatic, remarkable, significantly, outstanding, particularly, especially, in summary and in conclusion.* A question posed at the right time in your paper is great in capturing the attention of the reader and sustaining interest.

Limit conclusions

Avoid speculative statements by limiting any conclusions to findings in your study. A common mistake is to extrapolate findings done *in vitro* to make strong conclusions touching on animals or humans. Potential benefits in such studies should be stated conservatively.

Set realistic goals, avoid stagnation

Writing is a tiresome process due to the huge amounts of time and energy dedicated to the writing process. It is easy to excuse yourself before you finish writing important sections of your work. Setting achievable goals and milestones for a given task helps you focus. For example, you can decide that you are not reading your favorite morning newspaper until you finish writing the results section of the paper. If you achieve your goal, you will truly feel good, and have a sense of accomplishment. If this is done repeatedly over time, you will finish your paper without any delays. Sometimes, writers experience moments of hopelessness when they lack ideas, and the mind draws a blank. Sometimes, while writing you can be completely unable to grasp or explain a scientific phenomenon in your own data. It is advisable that you mark or make a mental note of that part in your document and continue writing other sections sympathetic to your brain. This allows you to keep that section in the back of your mind so that you can revisit it later. During the day, the brain is just clogged up with too many things. Note that while asleep, a resting brain creates room and incredible capacity in certain parts within itself, enabling it decipher difficult and complex tasks. Hence, as strange as it might sound, solutions to difficult unfinished sections of your paper can come as clearly as daylight in the deep of the night; more like a vision. If that happens, wake up, scribble whatever visions you had on something and crawl back to your bed and continue dreaming. Big problem solved and hopefully your partner will not be taking you to a psychiatrist the following day! However, if you are the one with such nocturnal habits, ensure to get rid of any children's toys on your floor before you sleep. Otherwise, a wrong move on them at night will send your newly found wisdom out the window within seconds. For other people, a coffee break or horse ride could do the trick. Do whatever works for you, but avoid stagnation and keep the momentum going as stated by Matthews (2008).

Editing

How many people see your first draft? Probably, other than yourself, only your advisor or research leader may see your first draft. It is a fact that the first draft will require extensive editing. Note that there is an improvement after every draft with editing. The paper must be edited until it is delightful to read. The simpler it gets to understand as you edit, the better the document. If you realize as the author, you are struggling to quickly comprehend your own document, then you have miserably failed in your writing and most likely, your experimental design is also flawed and your audience will be lost too. One of the unwritten rule is to have any such scientific document undergo a minimum of five revisions before a final document is presented for examination or review. Personally, I do at least ten revisions.

Secure your work

In this day and age, use of computers and other related gadgets is almost a necessity. As an author or a student undertaking scientific research, you may have come across horrendous tales of individuals who lost their entire grant application or thesis due to computer glitches. You must be critically aware that computers have a "strange mind of their own", and if you like, they do "fall ill" when most needed. Hence you must always save your work as often as possible and invest in backup facilities. Most computers have an option that automatically saves documents. Activate it, and send copies of your work to your secured email after every work session. Cloud computing provides additional avenues for document sharing and safety among many others.

The IMRAD structure

Note that even though scientific papers are varied in nature e.g. full research paper, concise paper, short paper, and short communications, they invariably adapt a similar structure i.e. the famous IMRAD structure (Introduction, Methods, Results, And Discussion). To quote Matthews (2008) in his book (Successful Scientific Writing), *"This so called IMRAD structure is not simply an arbitrary publication format, but rather a direct reflection of the process of scientific discovery"*. The IMRAD structure is meant to make scientific writing easier and consistent and should be followed.

The various structural parts common in scientific papers will be presented in their characteristic chronology in subsequent chapters (3 -11).

References

1. Alley, M. (1996) The craft of scientific writing, 3rd Ed. Prentice Hall NJ
2. Council of Science Editors (2006) Scientific Style and Format: The CSE Style Manual for Authors, Editors and Publishers. 8th Ed. New York: Cambridge University Press.
3. Day, R. (1998) How to write and publish a scientific paper. 5th Ed. Orynx press.
4. Goben, G. and Swan, J. (1990) The science of scientific writing. Am. Scientist 78:550-558.
5. Iverson, C.I. (1997) American Medical Association Manual of style: A Guide for Authors and Editors. 10th Ed. Baltimore MD: Lippincott, Williams and Wilkins.
6. Lai, J.C.K. (2002) Scientific Writing: A comprehensive introduction (Unpublished Guide).
7. Lebrun, J.L. (2007) Scientific Writing: A Reader and Writers Guide. World Scientific Publishing Co Pte Ltd, Singapore.
8. Matthews, J.R. and Matthews, R.W. (2008) Successful Scientific Writing: a step-by-step guide for the biological and medical sciences. 3rd Ed. Cambridge University Press, Cambridge, UK.
9. McMillan, V. (1988) Writing papers in the biological sciences. Bedford books. NY.
10. Publication Manual of the American psychological association (2001). 6th Ed. Washington DC: American Psychological Association.
11. Rubens, P. (2002) Science and Technical writing. A manual of style. 2nd Ed. Oxford: *Routledge study guides*.

Chapter 2

PLANNING IN SCIENTIFIC WRITING

Planning is the cornerstone of good and effective scientific writing. What does planning entail? It involves extensively thinking about what you want to write about and contextualizing it. This might involve brainstorming sessions. It is a process that can last for days, or even years. What follows is gathering information or literature that you want to use to support the writing process. It is at this planning stage that you decide the target journal, if you are writing a manuscript. The best practice is to decide on the Journal before you start the writing process. This is called Journal targeting. You must always target the paper to a particular journal based on a specific theme or interest, the best chance for acceptance for publication, and the potential to reach a wider readership/audience. After identifying a target journal, you need to read their most recent publications to familiarize yourself with their style and format. Be careful on where you publish your work. Publishing in Journals with poor reputation might haunt you for years as a scientist, and could result in lost opportunities along the way. Such approach in writing involving careful planning and targeting, is also applicable in writing of books or book chapters, research projects and thesis/dissertations.

It is not possible to write a good paper or thesis, if you have not read relevant literature in great detail, and have a good understanding of existing gaps in knowledge and any recent break through, if any. Extensive reading and marking of relevant literature and selection of core reference materials is critical. One of the unwritten rules is to ensure that you have read and cited papers or books published by those considered to be the scientific leaders in that field. These are respected and credible scientists in respective disciplines, whose contribution is conspicuous and outstanding. While it is not mandatory to read or cite such work, reviewers' and your audience might find it odd that you ignored distinguished scientists, whose work is relevant to your study. It is a probable occurrence that when some of the distinguished scientists receive a paper for peer review, the first thing they most likely do is flip to the reference section to check if they have been cited. Take note of that.

You must have a well thought out plan on how to handle loads of information in form of literature review that will be available online and in print. Particularly for print literature, marking and numbering is a great way to identify your sources. This will help when you need to go back to them. When using downloaded and printed material, make notes on the hard copies. For online sources, keep an online bookmark. You can also send any high value literature to your email. There are many ways to approach this, but what matters most is having a well laid down plan, and sticking with it until you finish the paper. Note that if you find a very good paper that is highly resourceful, it can assist in locating other equally useful literature to support your writing. Whatever you do, ensure that you use multiple sources of information such as research bibliographies, research registers, reference databases and citation indexes. These sources provide a rich source of vital literature to help your writing process. Note that failure to find certain literature does not mean it does not exist. To ensure your search does not draw blanks, carefully select your key words and use multiple search engines.

Some popular databases include *PubMed, general science index* and *web knowledge*, among many others. As you scour the internet for relevant literature, do so judiciously. You should be critically aware that not all information on the internet is credible. As a science student, avoid quoting Wikipedia as this is not considered a credible source. Obviously, citing such sources will not please your supervisor.

In this preparation phase, data analysis and organization is critical. You cannot start compiling your document until your data analysis is complete. This entails putting the data in a form that enables one to draw conclusions and makes it possible for potential readers in the same field to comprehend your story. Being aware of your most remarkable findings and hence the central message for your scientific write-up is very important. The readers must be able to pick this from the beginning to the end of your document. At this stage, the writer is now ready to pick up his/her computer and start hitting the key board.

References

1.   Alley, M. (1996) The craft of scientific writing, 3rd Ed. Prentice Hall NJ
2.   Day, R. (1998) How to write and publish a scientific paper. 5th Ed. Orynx press.
3.   Goben, G., and Swan, J. (1990) The science of scientific writing. Am. Scientist 78:550- 558.
4.   Lai, J.C.K. (2002) Scientific Writing: A comprehensive introduction (Unpublished Guide).
5.   McMillan, V. 1988. Writing papers in the biological sciences. Bedford books. NY.
6.   Matthews, J.R. and Matthews, R.W. (2008) Successful Scientific Writing: a step-by-step guide for the biological and medical sciences. 3rd Ed. Cambridge University Press, Cambridge, UK.

Chapter 3

WRITING A GOOD TITLE

The title is the first part of your document readers will come across. More or less, it is the face of your paper. By reading the title, some readers will decide whether the paper is worth their time. It should clearly reflect the contribution of the research. When choosing a title, think about what you want to communicate to the potential audience and the editorial team. The best title is the one that most clearly and directly relays the overall storyline or message of the paper. It is recommended that such title should be a short descriptive or declarative statement. The title should be catchy, unique and clear to both the expert and novice reader. It is possible to find some reports with typographical and punctuation errors within the title. Regardless of how outstanding the rest of the document is, this puts a damper in the works and creates a very negative first impression. Short titles are preferred, and the norm is to have not more than 100 characters (equivalent to about 10-12 words).

To come up with a good title, start with a draft that you feel is as close as possible to what you want. You can even have two or three versions. After writing the subsequent sections of the paper and attaining a reasonable draft, you can go back and re-look at your title and revise it accordingly. Most of the time you will find that you have a clearer picture of what you want in your title at this stage. The vital question always is whether the title captures the theme of your paper.

In some occasions, your title or sub-title might contain a word that is not familiar, that requires some explanation for the sake of your audience. It is recommended that the first sentence coming in the following section right after that title must contain the relevant explanation for that word. For research papers, such explanation will appear in the first line of the abstract.

Adhere to any word limits imposed on the title by your publisher. Use of key words at the beginning of your title will make it more effective. Some Journals specify that some key words are picked from the title. Whenever used, sub-titles should be written in a way that they are independent and not a duplication of the main title. Avoid use of trade names or brand names in the title of a scientific document. This applies to the summary/abstract.

Some of the words to avoid in your title include: *"study on…"*, *"Studies of…"*, *"The Implication of…"* Such titles do not reflect on any relevant information in the study, are not recommended and should be avoided. Examples of suitable titles are as follows:

a) *Catechins in tea extracts nullified neurotoxicity due to manganese in a mice model.*
b) *Kenyan purple tea anthocyanins ability to cross the blood brain barrier and reinforce brain antioxidant capacity in mice.*
c) *Pre-surgery counselling improved recovery from head trauma.*
d) *Coenzyme-Q_{10} prevented full blown splenomegaly and decreased melarsoprol-induced reactive encephalopathy.*

In most cases, such simple declarative or descriptive statements capture the intended central message without any ambiguity, and arouses interest in your audience. Note how the contribution of the research is placed upfront to create impact and interest. Use verbs and adjectives to give your title energy and make it exciting.

The following are examples of poorly structured titles:

a) *The study on the effect of caffeine in school children: a case study in Nakuru Kenya.*
b) *Investigations to determine the result of alcohol consumption on cholesterol among college students.*
c) *Does exercising in the morning affect sugar levels?*

Such titles do not provide any information on any observed results, and have zero message. Believe it or not, such titles appear on important scientific papers. Just ensure that your work does not become a source for such titles.

Be wary of titles with question marks as they might annoy your reviewers and the editorial board. Declarative or descriptive titles are preferred in most scientific journals; and they are really easy to write if you just take a moment and reflect on what your results are saying.

Some journals prefer that you capitalize each significant word in the title, while others prefer that you treat the title as a sentence. Refer to the specific requirements for your journal of interest in regard to capitalization.

Additional tips on how to write an effective title

a) Put any new findings or contribution at the beginning of the title.
b) Incorporate verbs and adjectives to the title to make it more exciting.
c) Use clear, specific keywords.
d) Be careful about choice of keywords used in the title as they will determine if your article will be located by online keyword searches.

If you follow these hints, your title will be impressive and will not result in lost opportunities or failing your thesis/dissertation examination.

Short/running title

It is a requirement by some Journals to have short/running titles. Such titles usually appear as the Journals header or footer. Note that the short/running title is also used for indexing purposes and hence a lot of thought should be put into its construction, just like the main title. Avoid duplication of the main title inform of your short title.

References

1. Day, R.A. (1994) How to write and publish a scientific paper. Cambridge University Press.
2. Lebrun, J.L. (2007) Scientific Writing: A Reader and Writers Guide. World Scientific Publishing Co Pte Ltd, Singapore.
3. Matthews, J.R. and Matthews, R.W. (2008) Successful Scientific Writing: a step-by-step guide for the biological and medical sciences. 3rd Ed. Cambridge University Press, Cambridge, UK.
4. Wilkinson, A.M. (1991) The scientists handbook for writing papers and dissertations, Prentice Hall, Englewood Cliffs, NJ.
5. Zeiger, M. (1991) Writing biomedical research papers, McGraw-Hill, New York.

Chapter 4

AUTHORSHIP

Authorship is an important aspect of scientific writing that is often neglected, and evidently, poorly addressed in many books. The pertinent questions is: who qualifies to be an author? Authorship is, and must always be tied to direct and significant intellectual input. The International Committee of Medical Journal Editors (2006) states that an author listed on the paper's title page should take responsibility for the scientific information in that paper. In other words, such a person must be able to defend that research if called upon to do so. There are a lot of disagreements and even legal cases emanating from authorship disputes. All these can be avoided if there is an initial agreement on the authorship order including those who will just be acknowledged or simply not mentioned at all.

Disputes mostly emanate from collaborative studies, and are partly due to dishonesty from disgraceful researchers keen to reap from where they have not sown. In some cases, legitimate intellectual contribution justifying inclusion as authors, is ignored by unscrupulous scientists, bent on taking all the credit resulting in disputes. Being honest and ethical in the practice of science can eliminate such disputes, so that more time is dedicated to research.

It is commonly accepted that the first and last authors are the most important. In most cases, the first author is the graduate student or post-doctoral scientist who did the experiments and majorly contributed in generation of the first draft. The last author is the head of the laboratory and most of the time, the originator and principal investigator of the research.

On some occasions, the head of the laboratory or principal investigator writes all manuscripts for research and data that has been generated by graduate students or other junior scientists in their laboratory. How will such students and scientists master their writing skills if they are not given an opportunity to write? Such laboratory leaders are not good mentors and can be described as misguided and even selfish.

Take into consideration that the sequence of authorship is important because it reflects the level of contribution to the research. It is advisable to agree on the sequence of authorship at the initial phase of the study. This will, without any doubt eliminate embarrassing controversies and disputes. Taking part exclusively in writing of the proposal or data collection does not automatically justify authorship.

In most institutions of higher learning, in addition to the principal investigator, two or more scientists with important expertize for the research in question are invited to constitute a supervisory committee for a graduate student. Such scientists do not qualify as authors unless they directly participated in the study design, execution of experiments or data analysis.

Across the world, even in the most advanced laboratories, it is not unusual to find that an equipment you need for a vital part of your study is not available in your laboratory. In such cases, the best approach is to involve another laboratory with such equipment in your study. The scientist in that laboratory with the equipment you need can be elevated into a collaborator if they bring expertise and even trains your laboratory personnel on how to use it. Such a collaborator qualifies to be included as an author. However, if you are lent an equipment without any technical support on how to use it, the best approach is to acknowledge that contribution.

Editors who provide editorial services should not be included as authors. Their inclusion would be unethical and undermines those who intellectually participated in the design, writing and execution of the experiment. I should state that in my practice as a scientist,

I have come across many instances where individuals who provide laboratory resource demand to be included in manuscripts. Providing a laboratory space/reagents or finances should not earn somebody authorship. Do not be mistaken, such support is indeed appreciated as it truly helps spur scientific discovery and optimal utilization of resources. However, these individuals or institutions may be mentioned in the acknowledgement section.

Before you include anybody in your paper as an author, you must inform them and get their approval. Failure to do this can cause a myriad of challenges that might include litigation. Note that some Journals require submission of a written statement usually in the letter to the editor, stating contributions made by each author.

What happens when two graduate students or post-doctoral fellows contribute equally to a paper and each deserves 1st authorship? For such cases, a statement declaring that the two are equal first authors should be included just after institutional affiliations. Below is an illustration on how it should be done:

Kenyan purple tea anthocyanins ability to cross the blood brain barrier and reinforce brain antioxidant capacity in mice.

Khalid Rashid[1,2,*], Francis N Wachira[2,3,*], *James Nyariki Nyabuga*[1], *Bernard Wanyonyi*[4], *Grace Murilla*[4], *Alfred Orina Isaac*[1,5]

[1]*Biochemistry and Molecular Biology Department, Egerton University, Egerton, Kenya.*
[2]*Tea and Health Department, Tea Research Foundation of Kenya, Kericho, Kenya.*
[3]*Programs Department, Association for Strengthening Agricultural Research in Eastern and Central Africa, Entebbe, Uganda.*
[4]*Pharmacology and Chemotherapy Division, Trypanosomiasis Research Centre, Kenya Agricultural Research Institute, Kikuyu, Kenya.*
[5]*Department of Pharmaceutical Sciences and Technology, Technical University of Kenya, Nairobi, Kenya.*
* - These authors contributed equally.

In related circumstances, it is possible to find an entire team in a laboratory where everybody contributed equally. In such a scenario, the alphabetical order can be used to determine the authorship order listed. A statement clarifying that all the authors contributed equally will be provided on the title page, just after institutional affiliations.

How do you handle technicians or research assistants? It is notable that some laboratories employ technicians who carry out the experiments on behalf of the scientist, but these technicians may not have participated in the design of the experiments or data analysis and write-up. They have simply been involved in running the experiments and generated data with no further input. In such instances, acknowledgement is the best option as is the case for those who provide funding.

When writing this section, always ensure that you clearly indicate the author(s) institutional affiliation(s). In some instances where the authors have changed work stations, include a heading saying "present address" which contains the necessary details.

The corresponding author is usually the Principal Investigator or head of a laboratory. Do not fail to provide the address, telephone, fax and email of the corresponding author.

References

1. Day, R.A. (1994) How to write and publish a scientific paper. Cambridge University Press.
2. Lai, J.C.K. (2002) Scientific Writing: A comprehensive introduction (Unpublished Guide).
3. The International Committee of Medical Journal Editors (2006)
4. Matthews, J.R. and Matthews, R.W. (2008) Successful Scientific Writing: a step-by-step guide for the biological and medical sciences. 3rd Ed. Cambridge University Press, Cambridge, UK.

Chapter 5

WRITING A GOOD ABSTRACT/SUMMARY

An abstract is a nonlinear reduction of your document such that some sections are ignored and others emphasized based on their strength. There is no requirement for proportional mention or inclusion in the abstract; whereas in a summary, all sections of the paper are presented or reduced in a manner that allows them to be equally or proportionately included. In fact, you can call a summary a miniaturized version of the whole paper. Usually a summary is much longer than an abstract. It is always important to make this distinction because different organizations will require one or the other. The word limits usually imposed on various write-ups determines the extent of reduction for either the abstract or summary. It is advisable but not mandatory to always write the summary/abstract last; because at this stage, the writer has in-depth understanding of the whole document. The take home message here is that the abstract and summary are different and should not be assumed to mean the same thing.

Abstract

For research articles, an abstract is a one paragraph reduction of your entire paper. It should never contain any information not covered in your document. Scientists who want to have a feel of your work without reading the entire paper may just read and rely on the abstract. Therefore, ensure that it is brief and informative and can stand on its own. Unless absolutely necessary, do not put citations in the abstract. Avoid use of words such as "I" and "we" in the abstract. Use the third person and past tense in this section. It is advisable to limit or completely avoid use of abbreviations, acronyms and symbols in the abstract. If they must be used, define them in the first mention in the abstract, as well as the first mention in the text. Limited statistical analysis information is acceptable but never a whole bunch of statistical numbers and symbols. Avoid pompous words that might not be familiar with your readers. You definitely do not want your readers to have a dictionary on their side, just to read your paper. Note that a preliminary decision to short-list an author for a grant may be made by just reading the abstract.

What is the function of an abstract and how is it organized? An abstract provides the glimpse of the entire paper and is adequate to encourage or discourage any further reading of the paper. It is not possible to overemphasize the importance of writing your opening statement for the abstract in a way that the problem you are studying is clear to your reader even if they are not experts. The abstract contains the opening remarks in the entire document, and must be concisely written. Period. Note that if by the third line of the opening remarks of the abstract the reader has not understood the purpose of your study, then you have failed in writing this section.

Generally, when writing the abstract, always make sure that you are brief and within any word limits provided. In most journal formats, the abstract is written as one paragraph with logical flow of information. It is best written using the author's own words.

What information is needed in the various parts of an abstract? The abstract for a scientific paper is organized into four main parts.

1. The first part of the abstract should give a brief background to the study and includes a clearly stated problem or question being investigated, and the purpose of your research. This must be written clearly and very briefly. On this first part, the hypothesis being tested must be clearly articulated. Here is an example on how to state your hypothesis in the abstract: "*… based on current literature and epidemiological*

studies, it is reasonable to hypothesize that continued rainfall will exacerbate malaria infections in Kisii, Nyanza province..."

2. The second part of the document, describes very briefly the most important aspects of the experimental design and the materials and methods used in the study. For example if you used a specific type of mice for your study, this is where you describe that very briefly. More or less, this part represents the entire methods section of your paper in a highly summarized version. If you mention any statistics and p values, always ensure that the data mean ± SD and sample size (n) are provided. According to the American Psychological Association (APA) guidelines, all numbers should be expressed numerically in the abstract. For the rest of the document, only numbers equal to or greater than 10 are expressed numerically. The exception is made with the numbers at the start of a sentence. It is possible that this requirement for numbers might be different in other Journals.

3. In the third part of your abstract, you should briefly state the most significant findings of your study. These are the results you want your audience to connect with, and are linked with the purpose of your study. Resist the temptation to present all results here. This is a mistake common with student thesis/dissertations. *Note that if you are writing a proposal, this part will not be there because you do not have any results to report.*

4. The fourth and last part of the abstract must briefly and clearly state the implications or importance of the research findings.

Note that a clear understanding of the main parts of the abstract as presented above will enable you to plan your information and write it easily and succinctly without any problems.

A very keen student will be interested to know that when astute scientists want to get to the gist of your study without reading the entire paper, they will usually read the abstract and the first and last paragraph of your introduction. You must take note of this! If a lot of thought has not gone into writing this parts, your audience will lose interest in your work and not read it further.

Key words

As a young scientist or graduate student, it is important to understand the importance of careful selection of key words. Why should anybody worry about choice of key words?

a) Key words are used for indexing and inclusion in other relevant data bases that cite or list the journals and articles your target has.

b) Ultimately, your choice of key words will impact on the online presence of your articles. If correctly selected, key words can boost your citation rates.

c) Key words selected from published and related work, will tremendously improve the article retrieval online.

d) The number of citations recorded from publications as a result of a specific project are indicative of the success of that research.

e) Note that if you do not provide key words as requested, somebody in the editorial office will select them for you using statements from your title. This is not a good practice because the selection could be inappropriate for the paper.

In choice of key words, be careful to ensure that they reflect the theme of your paper. It is recommended that key words are written last due to the thorough understanding of the work at this stage.

References

1. Bem, D.J. (1987) Writing the empirical Journal Article. In M.P. Zanna and J.M. Darley (Eds.). The complete academic: A practical guide for the beginning social scientist (pp. 171-201). New York: Random House.
2. Matthews, J.R. and Matthews, R.W. (2008) Successful Scientific Writing: a step-by-step guide for the biological and medical sciences. 3rd Ed. Cambridge University Press, Cambridge, UK.
3. Mugenda, O.M. and Mugenda A.G. (2003) Research Methods: Quantitative and qualitative approaches. Acts Press.
4. Zeiger, M. (1991) Writing biomedical research papers, McGraw-Hill, New York.

Chapter 6

THE INTRODUCTION

The introduction sets the stage for the entire study. This section introduces the problem or challenge being investigated and the hypothesis behind it. This is an important section of the document that must be written without any ambiguities whatsoever. A well written introduction should:

a) State the problem clearly.
b) Create an elaborate build-up to the problem leading to the study.
c) Relate your study within the latest breakthroughs in the area, while targeting your audience.
d) Stimulate interest in your topic.

For ease of writing, this section should be organized as outlined below:

1) The first portion of the introduction should clearly articulate the problem studied. Any relevant literature cited should clearly identify the existing deficits in knowledge that your study aims to fill and must be linked to the purpose of your research. When you state any such deficiencies, indicate how your study wholly or partly addresses them. It is at this point that the author should write in a manner that reaches out to the target audience and in a concise way, state the hypothesis of the study. This will of course be presented with appropriate support from existing and current literature. On this section, the question(s) you want to answer are presented clearly and will represent the core theme of your work that must permeate all other subsequent sections.

2) The second part should clearly state the importance or implications of the study such that someone can read this part and have an understanding of what the research is all about. Usually, information on this part is consistent with that in the last paragraph of the abstract.

Both part one and two of the introduction as presented above may consist of more than one paragraph.

In some formats, you will have a section called introduction (as presented above) and a separate one called literature review. In such a case, the literature review will take a similar format as described for the introduction, but will be more detailed and extensive; and might have sub-headings. Such a format with a separate section for literature review is common with dissertation/thesis/research projects and review papers; and allows the student to explore the existing literature deeply, helping support certain arguments and conclusions. Note that in review articles, the literature review and theoretical information is very detailed and extensive, and constitutes the main body of the paper.

Take note that you do not have to cite all the literature you have studied. You only include literature review relevant to your research. Remember that the purpose of writing a scientific paper is to add or contribute new information to the existing body of knowledge. Hence the need to ensure that the review of old literature does not overshadow your own findings. If you feel some studies have more compelling findings that your audience will appreciate, then cite them.

If by any chance you find interesting images, pictures or illustrations that will help communicate to your readers, use them. Just ensure that you write to the corresponding author of the paper with the image of interest and seek their permission to use it. Most of the time, you will get a positive response for such requests. With the author's authority, you can now use that image with a caption saying for example: *"this image used with permission from Isaac et al., 2013"*.

References

1. Creswell, J.W. (2009) Research Design: qualitative, quantitative and mixed methods approaches. Third Edition. SAGE publications Ltd, Thousand Oaks California.

2. Lai, J.C.K. (2002) Scientific Writing: A comprehensive introduction (Unpublished Guide).

3. Matthews, J.R. and Matthews, R.W. (2008) Successful Scientific Writing: a step-by-step guide for the biological and medical sciences. 3rd Ed. Cambridge University Press, Cambridge, UK.

4. Mugenda, O.M. and Mugenda A.G. (2003) Research Methods: Quantitative and qualitative approaches. Acts Press.

5. The Basics of Scientific Writing in APA style: Manual of The American Psychological Association (6th Ed.) 2010.

6. Wilkinson, A.M. (1991). The Scientists handbook for writing papers, dissertations. Eaglewood Cliffs, NJ: Prentice Hall.

Chapter 7

THE MATERIALS AND METHODS

Awell written materials and methods section provides enough details and references in adequate details to enable other scientists repeat your experiment(s). A well written methods section must fulfil this requirement.

Paying special attention to the questions below will help you tackle this section satisfactorily.

a) What did you do to answer questions raised in the study?
b) How was it done?
c) Why was a specific procedure chosen for your study and not others?
d) How did you analyze your data? What methods including statistical ones did you utilize to organize and analyze your data?

This section will include the following details:

1) The Materials sub-section usually consisting of Chemicals or other substances used in the study like sodium chloride or antibodies. You must always indicate the source of such materials i.e. manufacturer, city and country of origin. Materials also include any special items like media, tissues, cell lines, and equipment. In addition, use of any specialized animal models or humans in your study will be described here. When animals or humans are involved in any study, you must declare adherence and ethical clearance for animal/human use from relevant institutions in your area.

2) The Methods sub-section. This section will consist of additional sub-sections namely:

i. Mode of preparation e.g. whole tissue, homogenates, cell lysates etc.

ii. The step by step protocols. A good example here are the protocols for polymerase chain reaction (PCR) or Western blots or ELISA tests. Sometimes a lot of details for such protocols will be cut out because other scientists will be well versed with such and can make references from other previous work on the same. However, it is important to give detailed information on specialized bio-reagents like anti-bodies or restriction enzymes and quantities used to enable other scientists repeat the study. It is very important to indicate any modifications on published protocols used in your study. As stated by Benson and Boege (2002), publically funded research should be conducted in accordance with Good Laboratory Practice (GLP) guidelines, which require availability of standard operating procedures in the laboratory involved.

iii. The rationale for choosing specific methods and procedures should be stated. This is important because other methods other than the one used could be applicable.

iv. Any assumptions and limitations to the methodologies used should be stated.

v. Information on methods used to collect, organize and analyze data.

vi. It is recommended that international units of measurements (SI units) are used. However, whenever that is not possible, it is important to provide additional details and equivalence to known SI units.

Remember that you should use species name or common names of animals/organisms utilized in the methods section.

Avoid the following:

a) Do not write all the details of what was done, but rather provide sufficient information for other scientists to follow your work, and where needed, be able to repeat your experiments without any difficult.

b) Do not hide information in this section with the intention of making it difficult for competitors to repeat your experiment(s). Such conduct would clearly violate ethics in science.

References

1. Benson, B.W. and Boege, S. (2002) Handbook of Good Laboratory Practices. Bristol, PA: Hemisphere Publishing.

2. Lai, J.C.K. (2002) Scientific Writing: A comprehensive introduction (Unpublished Guide).

3. Matthews, J.R. and Matthews, R.W. (2008) Successful Scientific Writing: a step-by-step guide for the biological and medical sciences. 3rd Ed. Cambridge University Press, Cambridge, UK.

4. Mugenda, O.M and Mugenda A.G. (2003) Research Methods: Quantitative and qualitative approaches. Acts Press.

Chapter 8

HOW TO PRESENT RESULTS

This section is very important because the results presented are ultimately aimed at responding to your hypothesis and questions you intended to answer. All other sections of your paper rely on information or observations made in the result section. Data collected and analyzed using various tools and presented in various forms such as figures and tables is what makes up the results. Results are therefore conclusions drawn by analyzing data, and include specific statements.

It is important to exclude findings not really relevant to your work, while taking care not to exclude legitimate results because they appear contrary to your hypothesis. For example:

Suppose you are studying the impact of tea polyphenols on brain antioxidant capacity in a mice model; and in the process notice that the polyphenols induce weight gain in mice. Will you report the effect of polyphenols on weight in your paper? I do not think so. Others might have a differing opinion. Note that although you noted weight gain, you cannot explain it, at least until further studies have been done. Moreover, while it is important to take note of such observations, they should not distract from the focus of the study. These observations are not oblivious to the fact that such unintended observations, have in the past resulted in tremendous scientific discovery and innovation. A good example is the discovery of Penicillin.

You simply do not have anything to write about without data that has been carefully analyzed and converted into results. You are encouraged to have results from your data analysis in front of you at all times as you write this part, to help you maintain focus and organization. At this moment, you should have organized your results logically, in order of their importance. Most result sections are presented in sub-sections/sub-titles to give them a clear focus, sequence and order. You must arrange the results in such a way that you clearly know, which information goes to each sub-section before starting to write the paper.

It is advisable to present the results in such a way that the most significant findings with regard to the central question of your study or theme come first. Astute scientists use this to their advantage, and in my view should be the preferred approach. Other schools of thought might argue that you instead should present the results in concert with the order listed at the objectives section. This arrangement is then maintained throughout the document. As stated early, this approach is not popular in scientific practice because your audience will be really struggling to see the significance of your work, which might be truly present but buried at the end of your paper. To fully appreciate the impact of presenting important findings first, think of an editor of a popular newspaper putting the juiciest story on the back page! That would not make any commercial sense and will be insensitive to the readers.

The temptation to falsify data has always been present in scientific practice, especially in statistical analyses. It is important to always present data truthfully. Be aware that many individuals, some prominent persons in society have had their doctoral degrees withdrawn many years after graduation due to scientific malpractice. One of the consequences of data manipulation is that other scientists are not able to replicate findings in such research. Inability to replicate results raises a red flag for possible data fraud. Note that greater confidence in scientific research is earned if the scientist has a good reputation and is a person of great integrity.

When reporting statistical studies, the emphasis should be on the science and not provision of detailed and complex mathematical assumptions and statistical details of the tests done. What is required of a biology/biomedical/medical student or scientist in regard to statistics is a simple statement on the chosen test, the mean and standard deviation and the probability level or p value, and sample size (n). One common error among young scientists is describing results not shown to be statistically significant as: *"nearly significant"* or *"results showed a remarkable increase, though statistically not significant"*. If you did correct statistical analysis with the correct methods and tools and got no significance difference, then the results must be reported so. Equally common is the use of the word "significant" to refer to changes in results where statistical analysis was not done. Take note of these errors and avoid them.

It is common to present results in form of tables, figures and pictures. As it will be apparent to you as a researcher, a good picture tells the story much more directly and dramatically than pages of verbiage. Note that not all data is appropriate for presentation in the form of a table or figure. Simple data is best presented in the text in one sentence with appropriate statistics put in parenthesis. However, visual aids in form of tables, figures and pictures constitute powerful tools for data presentation in the result section. The visual aids must be informative and well done. Remember that use of visual aids can help emphasize scientific evidence in the study. Ensure that your visual aids in whatever form, helps communicate scientific information more clearly. If they do not meet this threshold, do not use them. A figure occupying a quarter of a page, can adequately communicate two pages of information written in text. Use the figures diligently to make your paper short and concise. Note that some of your audience might prefer to have a brief look at the visual aids before they read the paper. Figures, when presented well, in a logical and simple manner, capture and sustain the interest of your readers, and help them synthesize information faster as described by Lebrun (2007); who correctly refers to visual aids as the fast food of the brain.

Note that you cannot present the same data in the table format and figures. You should choose one. Usually, it is best to choose what will communicate most effectively to the readers. For example, if you are reporting dose response studies, a figure in form of a line graph will be the best option because it will clearly demonstrate any trends and interactions if you have more than one variable. However, if you are engaged in an epidemiological study to determine how many people are infected with malaria in a certain area, you might best present that in a table. For comparative studies, bar graphs or histograms will be preferred. Graphs are important when drawing comparisons, as they are easily understood as compared to lengthy written explanations.

It is preferable to always present bar graphs/histograms using patterns and not different color schemes. Use of patterns ensures that the differences in the various treatments as presented in your bar graphs are discernible even when viewed or printed in black and white. The figures below, one presented with patterns (Fig.1) and the second a color scheme (Fig.2), illustrate my argument. It is possible that other schools of thought might have a differing opinion.

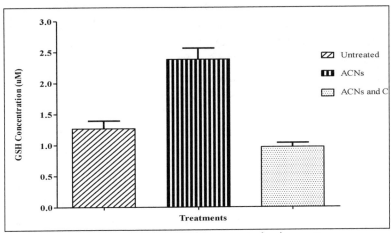

Figure 1: **Preferred style:** Bar graphs presented using patterns.
Even if this is printed in black/grey scale, the bar graphs will be distinctly clear. Used with permission from Isaac and Khalid (2013).

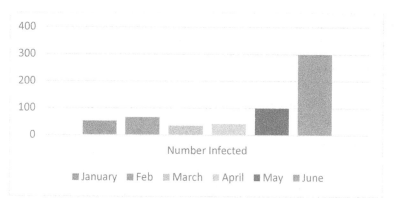

Figure 2: Malaria infection rates in Kisii district, January - June, 2012.
Style not preferred: *Bar graphs are presented using color schemes. Such colors will not be discernible if this page is printed in black and white or grey scale.*

Note that something else is wrong with figure 2. It does not have error bars like figure one. For scientific or academic papers where data has been subjected to statistical analyses, error bars should be inserted appropriately.

Pay particular attention to the size of your figures. Consider at least a ratio of 2 vertical units to 3 horizontal units for your figures as outlined by Matthew (2008). If this is done, the figures will easily fit into allocated space in your Journal presentation format.

Simplicity is key when dealing with figures in formal scientific writing. Avoid multi-colored 2D/3D fancy histograms or graphs.

In your final typescript, do not include statements such as; "as shown in the figure above", because pages and positions of figures can change after typesetting.

Data that is diligently presented in a table format communicates the intended information at a glance, even to individuals in another field of study. Unless specified by your journal, do not have grids, or vertical lines in your tables. However, three horizontal lines are preferred as shown in table 1.

Table1: An illustration on how to present a table

Values (means ± SEM) of relapse period in days in mice infected with *Trypanosoma brucei rhodesiense* and employed for PTRE studies.

Mice group	Mean relapse period (days) ± SEM	Range
PTRE controls	23.33 ± 1.453^a	21–26
PTRE and Co-Q_{10}	28.20 ± 1.770^a	25–35
PTRE and ACNs	29.75 ± 2.500^a	25–35
PTRE, ACNs and Co-Q_{10}	29.20 ± 1.855^a	25–35

Treatments marked with the same letters are not significantly different at $p < 0.05$.

Used with permission from Isaac and Khalid (2014).

For tables, it is easy to compare items listed going down. This should be the preferred format. Generally, the independent variables are better presented in rows horizontally and the dependent variables in columns vertically. This approach produces concise tables, which are easy to analyze and comprehend. With numerical data, do not use fractions; decimals are preferred. Sometimes, there is too much information in one table making it complex and potentially difficult to understand. Consider splitting such a table into two.

For line graphs, simplicity is very important. Limit the number of lines per graph. Once again, distinguish different lines within a graph by using symbols or patterns rather than color schemes. As clarified earlier, color schemes loose clarity when printed or viewed in black and white. If there are comparisons between graphs, ensure that the scales are the same.

Important aspects to note about the result section:

1. This section is reported in past tense.
2. Scientists should only report experiments that worked. Failed experiments can be reported if such information would be of scientific value.
3. The data presented here must be accurate and consistent. It is important to state the statistical

treatments used and hypothesis tested. Whenever you use the words "markedly or significantly increased or decreased", as the case might be, indicate the supporting statistics and p values and state that in quantitative terms. Take note that the words "significantly changed", should be restricted to presentation of statistical significance and must be accompanied with appropriate p values.

4. It is advisable to use sub-headings to enable your readers determine movement from one area of focus/experiment to the other.

5. For every result presented, there is usually a corresponding methods section.

6. It is recommended that you mention the species and materials used for the study at least once in the result section, describing how the experiments were done.

7. Not all the data generated will be included in this section. You will carefully select data that is relevant and answers the questions that your study intended to address.

8. For this section, you only present the results, and should not state their implications or discuss them. Simply state what your data shows without any explanations. There are exceptions to this rule especially in thesis/dissertations where the result section may be merged with the discussion.

Caption/Legend for tables and figures

A caption is a descriptive title or headline, usually consisting of a few words describing the contents of a graphical illustration or picture. The legend on the other hand, consists of information describing the figure. It provides more details for abbreviations and any symbols used. Whenever the caption and legend are presented one after the other, the caption ends with a period. For clarity, an illustration is provided in figure 3.

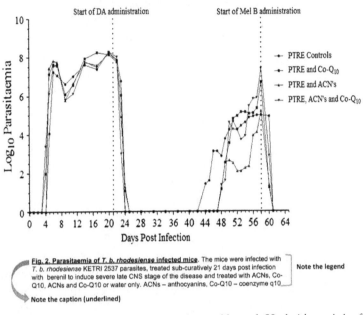

Figure 3: An illustration depicting a caption and legend. *Used with permission from Isaac and Khalid (2014).*

A caption and legend describes what a figure or table is showing in brief but adequate detail for the reader to understand without making references to other sections of the paper. Simply put, the table/figure should explain itself with a well written caption and legend. The caption and legend consists of a number e.g. figure 1(a), a short title and a brief description. Note that for figures, the legend which includes the title, is normally put below the figure as shown in figure 3. Ensure that for a research paper, no legend is included in the figure itself. For tables, the title or caption is usually above the table. However the more detailed legend may appear below it. Figures are usually self-explanatory, and you only need to state their vital features in the legend, without delving into lengthy statements describing relationships shown. The reader is able to see that.

For a research paper, legends for all the figures are usually placed in a separate page at the end of the document, while for tables, the titles are placed on top of each table. Such a title provides brief, but adequate detail to be understood without reading the text. Always remember to include sample size details (n) in the table and figure legends as appropriate.

The caption must be close to the picture, and readers should not put too much effort looking for it. A well placed caption below a picture helps the reader and can create interest in reading the rest of the paper. A picture without a caption is a nuisance to the reader, and can discourage any further reading. Due to the importance of the caption and legend, a lot of thought must be put in their construction. In the same spirit, ensure that information provided in the caption/legend will assist your readers, and not confuse them.

In summary, visual aids are an important component of the result section, since they help communicate the storyline or theme to the audience.

A more detailed description on how to present figures and tables can be found on Matthews and Matthews (2008).

References

1. Day, R.A. (1994) How to write and publish a scientific paper. Cambridge University Press.
2. Lai, J.C.K. (2002) Scientific Writing: A comprehensive introduction (Unpublished Guide).
3. Mathews, J.R., Bowen, J.M., and Mathews R.W. (1996) Successful Scientific Writing: A step by step Guide for Biomedical Scientists, Cambridge University, Cambridge U.K.
4. Matthews, J.R. and Matthews, R.W. (2008) Successful Scientific Writing: a step-by-step guide for the biological

and medical sciences. 3rd Ed. Cambridge University Press, Cambridge, UK.

5. Mugenda, O.M. and Mugenda, A.G. (2003) Research Methods: Quantitative and qualitative approaches. Acts Press.

6. Wilkinson, A.M. (1991). The scientists Handbook for writing papers and dissertations, Prentice Hall, Englewood Cliffs, NJ.

7. Zeiger, M (1991) Writing biomedical research papers, McGraw-Hill, New York.

Chapter 9

DISCUSSION AND CONCLUSION

The discussion section provides the author with an opportunity to make comparisons of their finding with those from other scientists through existing literature. It is always important to discuss literature that agrees with your findings as well as that which presents contrary findings.

A common mistake mostly observed with graduate students, is trying to explain and justify all the observations noted in their study. This is not necessary and can make your document undesirably long. While it is important to discuss all results to a certain extent, there should be more focus on what you consider your most significant findings, relative to the purpose of your research.

Before you start writing this section, you should have in mind some of these questions:

a) What is my most significant finding in this study?
b) Did the results support the hypothesis?
c) Which knowledge gaps are addressed by my findings?
d) What are the key limitations of my study?
e) Are there studies that corroborate my findings? Are there other studies with contrary findings? *Offer explanations for any differences between your findings and prior studies.*
f) If you were to pursue the study further, which direction will it follow?

If you have clear answers to these questions, it will be relatively easy to write this section.

Note that just like the other sections presented in this guide, the discussion section is also structured as follows:

1. In the opening remarks of the discussion, you must always state your most significant finding(s) and relate them to the hypothesis of your research or the pertinent questions raised by your study in the introduction. The latter part of this section should present other findings in the study; and allows the authors to compare and contrast their findings with existing studies.

2. The second part of the discussion section should address broader issues raised by the study and the implications of the findings. The author should discuss any strengths and limitations of the study, and how that affects conclusions drawn from it. In addition, the author should shed some light, in their view, the direction such a study might take in future to address any gaps or deficits. Bear in mind that such suggestions for future studies should be made with the benefit of hindsight for previous studies, critically aware of any gaps in literature. As you conclude this part, ensure that the reader of your research findings has a take home message, and is best written in your own words without references to already published literature.

3. Whenever appropriate, include recommendations as the last part of this section.

Points to note:

a) Be careful about extrapolation of results; for example, findings in mice to humans.

b) Respect other people's work and do not disparage their research, especially by comparing it to yours.

c) It is ok to admit any unexplained anomalies in your data.

d) Use present tense for answer(s).

e) Use strong verbs for conclusions and applications e.g.

> *"Purple tea anthocyanins blocked neurotoxicity due to manganese".*

f) Use weak verbs for implications and speculations e.g.

> *"Findings suggest that purple tea anthocyanins may lower manganese neurotoxicity".*

> *"Findings imply purple tea anthocyanins may assuage manganese neurotoxicity".*

Note the difference between the statements in e and f above. One is stated with finality (e), and the others (f) have left room for further determinations. This is an important stylistic method you should learn.

g) At all times, you should avoid speculative statements and limit your conclusions to the findings in your study.

h) For some scientific write-ups such as short communications, the discussion section is merged with the result section. This is also true for some Universities, which may require that students submitting their dissertations for examination merge the result section with the discussion. Note that whether merged or not, the purpose and structure of the respective sections stands as previously presented.

i) In most thesis/dissertations, students include a lot of content hence resulting in very many pages in this section as a way of trying to explain the results. As per Hailman and Strier (1997), this section should not be the longest part of your paper.

References

1) The Basics of Scientific Writing in APA style: Manual of The American Psychological Association (6th Ed.) 2010.
2) Hailman, J.P. and Strier, K.P. (1997) Planning, Proposing, and Presenting Science Effectively. Cambridge UK: Cambridge University Press.
3) Matthews, J.R. and Matthews, R.W. (2008) Successful Scientific Writing: a step-by-step guide for the biological and medical sciences. 3rd Ed. Cambridge University Press, Cambridge, UK.

Chapter 10

THE ACKNOWLEDGEMENT

In most journal formats, this section is strategically put immediately after the discussion. It provides an important tool for the scientist to thank all those who supported their research, mostly financially and technically. Describe such support very briefly and indicate unique grant numbers. All sources of financial support, however meagre, should be reported here. In some Journal formats, funding bodies can be acknowledged on the title page. Conform to your specific journal format. Note that this section is not meant for thanking family or friends.

It is always required that you inform those you intend to include on this section in advance because some Journals require their written authority.

Chapter 11

THE REFERENCES

It is not acceptable to pick sentences from somebody's work and insert them in your text word by word. The acceptable way is to restate the authors' ideas in your own words and then provide the citation appropriately. There is no other way about it. As a scientific writer, you must endeavor to avoid plagiarism by ensuring that you always acknowledge or provide citations without failure.

Do not insert a reference on your list, picked directly from other articles. You can imagine the consequence of such conduct if the other author had cited that reference wrongly. Adhere to any reference number limits by your Journals.

Under the reference section, journals/books/reports cited in the body of the paper are listed in a form specified by the journal or institution in which you intend to publish your work. How the citations are done in the text is also determined by the format specified by your publisher. I will give a few examples to illustrate my statement:

In the text, many journals require that the authors insert more than two citations thus: (Isaac et al., 2006; Isaac et al., 2007). The references are listed in a chronological order in the text, starting with the early publication as shown. These are listed in the reference section at the end in a chronological order:

1. **Isaac, A. O.,** *Dukhande, V.V., Lai, J.C. (2006). Metabolic and antioxidant system alterations in an astrocytoma cell line challenged with mitochondrial DNA deletion. Neurochem Res. 32(11):1906-18*

2. **Isaac, A.O.**, *Kawikova, I., Bothwell, A.L., Daniels, C.K., Lai, J.C. (2007) Manganese treatment modulates the expression of peroxisome proliferator-activated receptors in astrocytoma and neuroblastoma cells. Neurochem Res. 31(11):1305-16*

This style is recommended by the Council of Biology Editors Style Manual, 5th Ed. (1993) and the American Chemical Society style guide (1986).

In another style recommended by the Council of Biology Editors Style Manual, 5th Ed. (1993) and the American Chemical Society Style Guide (1986), references are numbered as per their first mention in the text. Simply put, a number is assigned to the reference and is placed in parenthesis. Subsequently such references are then listed in a numerical order in the reference section.

Example:
Citation In the text: '*Mitochondrial DNA damage interferes with cell signaling pathways (1, 2)*'.
In the reference list:

1. **Isaac, A.O.**, *Dukhande, V.V., Lai, J.C. (2006). Metabolic and antioxidant system alterations in an astrocytoma cell line challenged with mitochondrial DNA deletion. Neurochem Res. 32(11):1906-18.*
2. **Isaac, A.O.**, *Kawikova, I., Bothwell, A.L., Daniels, C.K., Lai, J.C. (2007) Manganese treatment modulates the expression of peroxisome proliferator-activated receptors in astrocytoma and neuroblastoma cells. Neurochem Res. 31(11):1305-16.*

Note that regardless of the format specified by your target journal, it is recommended that for your first draft, you use the Harvard system (uses the name and year) for citations instead of using consecutive numbers (Vancouver style). This is purely meant to make your organization task simple because it is easy to recognize names when making your final reference list.

Once you have chosen the format to use as specified by your Journal of choice, ensure that you maintain consistency in your method of citation throughout the text and the reference list at the end. It is quite common to find a very well written paper, in which this part has been completely neglected. One of the most common mistakes is citing references and then forgetting to list them in the reference list. Another equally common error is mixing up the order as cited in the text, and therefore having the papers cited assigned a wrong number at the reference list. Ultimately, this error will mean that authors will be assigned sections of the text that is not related to their publication and research. Definitely, you do not want editors or reviewers to come across such errors in your paper.

As stated earlier, various journals or institutions have specific formats for doing the reference section. Make it a practice to use the most recent publication from your target journal for use a practical guide when formatting your references. This is equally useful when writing other sections of your paper.

References

1. Council of Biology Editors Style Manual Committee (1994) Scientific Style and Format: The CBE Manual for Authors, Editors, and Publishers, 6th Ed., Cambridge and New York.
2. Dodd, J.S. (1986) The ACS Style Guide: A Manual for Authors and Editors, American Chemical Society, Washington DC.
3. Isaac, A.O., Dukhande, V.V., and Lai, J.C. (2007) Metabolic and antioxidant system alterations in an astrocytoma cell line challenged with mitochondrial DNA deletion. Neurochem Res. 32(11):1906-18
4. Isaac, A.O., Kawikova, I.B., Daniels, C.K., Lai, J.C. (2006) Manganese treatment modulates the expression of peroxisome proliferator-activated receptors in astrocytoma and neuroblastoma cells. Neurochem Res. 31(11):1305-16.
5. Lai, J.C.K. (2002) Scientific Writing: A comprehensive introduction *(Unpublished Guide)*.
6. Matthews, J.R. and Matthews, R.W. (2008) Successful Scientific Writing: a step-by-step guide for the biological and medical sciences. 3rd Ed. Cambridge University Press, Cambridge, UK.

Chapter 12

OTHER SECTIONS IN THESIS, RESEARCH PROJECTS AND PROPOSALS

Objectives

Usually in journal articles, the aims or objectives of the study are integrated into the introduction section and more or less represent the theme of the study. However, in thesis/dissertation/Research projects/proposals, this usually constitutes a separate and distinct sub-section.

The objectives must meet a certain common criteria thus:

a) The objectives are derived from the purpose of the research.
b) They have to be stated clearly, and very briefly (preferably, not more than one line).
c) They must be realistic, testable and time bound.
d) An objective should not have more than one actionable verb.

Objectives are crucial in that they explicitly determine the type of questions to be asked. Any such questions must address specific objectives provided. In addition, the objectives determine the methods to be used, data collection tools and analysis procedures to be deployed. In essence, they provide a sort of guideline for the research.

Expected outputs

What is the author expected to capture in this section? Note that expected outputs should be potential benefits accruing from your research, and should be tangible and measurable. Some publishers or institutions may prefer to have time limits.

Expected outputs are linked to the objectives. Ideally, each objective should have its own expected output(s). This is usually not the case because, more often than not, results for certain objectives are not strong enough to qualify as solid outputs.

In writing the expected output, do not use more than one actionable verb within one output.

Problem statement

This section should be written using concise statements to clearly state the problem the study is investigating or the purpose of the research. In some formats, this section is called problem statement, while others call it statement of the problem. Before writing the problem statement, ensure you have a thorough understanding of the broader area being studied. This knowledge can be acquired through literature review, observations or discussions with those with deep knowledge in the subject. This section must capture the storyline or theme of your research. You might ask yourself, "what is the problem my study is trying to solve?" If you have a clear answer to this question, then you will find it easy to write this section.

The last paragraph for this section must state the implications or importance of the study in question. Writers are encouraged to write this section using their own words, without any references or citations. It must be as brief as possible, usually less than 150 words or about half a page (double space, font 12).

In summary, the following events may guide you to formulate a superb problem statement:

a) Think about the question being investigated in a broad way.
b) Seek and study relevant literature for any recent breakthroughs or deficits/gaps in knowledge. Take note that in as much as we appreciate current literature, you must include initial or old, but important findings in the area.

c) Carry out interviews or have discussions with colleagues or experts in the area.

d) With hindsight on the purpose of the research, write your problem statement as clearly as possible using your own words.

Hypothesis

Some will define hypothesis as the anticipated or result of the study. A hypothesis is based on existing knowledge in that field. Scientists are advised to give a lot of time and careful thought in formulating their hypothesis because it forms the focal point of the research. Basically, your study tests your hypothesis using conclusions drawn from your data analysis. The hypothesis must be testable within a specified time, and is limited to the research being done. One of the roles of the hypothesis is to guide the scientist by delimiting the area of research and keeping them focused.

It is always important to ensure that the hypothesis is consistent with the purpose of your research. One study can have more than one hypothesis if it has more than one variable.

It is possible to do a research without a hypothesis. This scenario is possible when doing explorative research without any prior knowledge or scanty previous research. In this type of study, the objectives will suffice to guide the research.

Justification

If you have a clearly stated problem and provide compelling information justifying your study, then you may consider this section complete. This section cites compelling literature and other relevant statistics providing critical information on the importance of addressing a certain challenge, deficit, or knowledge gaps. It is important to include numerical information for impact *e.g. 500 million people die from malaria every year.* However, people may argue that the previous statement is misleading because some scientists study basic science for the sake of knowledge, without any intention for direct benefits in the short and even long term.

Ensure that this section is short (usually not more than one page), and use compelling language to reiterate the need to pursue your study.

Many students have difficulty separating information going into the problem statement and justification. Note that in the justification section, you have more wiggle room to cite literature and relevant statistics to build your case for the need for the study. On the other hand, in the problem statement, you simply state the problem after digesting relevant literature and should have minimal or no citations.

Can the justification, in a way, be equated to the rationale of the study? Yes, this is possible. Just ensure that your audience will understand and appreciate what you have covered in your study, and the reasons why you did it.

Literature review

Some scientific writings have an additional section other than the introduction called the literature review. Some of these documents include reviews, research proposals and thesis/dissertation. In my view, a separate section on literature review is intended to give the author an opportunity to demonstrate deep understanding of existing and relevant literature. Just like the introduction section, the author must focus their attention on relevant literature, including any gaps in knowledge. A common mistake in this section is loss of focus and presenting literature that might be related to the current subject but irrelevant to the study at hand. *For example, if a given study is investigating cell signaling aberrations in neuroblastoma, should it have extensive literature on leukemia?* This should not be the case. Many writers veer off their path and discuss unnecessary topics.

For clarity and logical flow of information in this section, it is acceptable to have subsections focusing on various aspects of the study. The flow of information must be consistent with the order in the objectives or methods sections.

References

1. Bem, D.J. (1987). Writing the empirical Journal Article. In M.P. Zanna and J.M. Darley (Eds.). The complete academic: A practical guide for the beginning social scientist (pp. 171-201). New York: Random House.
2. Creswell, J.W. (2009) Research Design: qualitative, quantitative and mixed methods approaches. Third Edition. SAGE publications Ltd, Thousand Oaks California.
3. Kothari, C.R. (2006) Research Methodology: Methods and techniques. *New Age International (P) Limited, Publishers.*
4. Maxwell, J.A. (2005). Qualitative research design: An interactive Approach (2nd Ed). Thousand Oaks, CA: SAGE.
5. Mugenda, O.M and Mugenda, A.G. (2003) Research Methods: Quantitative and qualitative approaches. Acts Press.
6. Ogechi, S.N. (2013). Research Methods: Question and Answer Revision Book.

Chapter 13

EDITING AFTER THE FIRST DRAFT

B y this stage, you have done a lot of work and having a complete draft is in itself a big milestone. For many writers, even in subjects outside science, getting a complete draft boosts confidence, providing great motivation to finish the paper. As mentioned earlier in the planning stage, the first draft is not shared with colleagues because it is still raw and requires a lot of editing. Hence the following actions are recommended in the order presented:

a) Check the entire draft to ensure that all sections have been included as per the required format. Fill up any gaps present.

b) Read your first complete draft aloud in a manner simulating a reader other than yourself. This act puts you in the shoes of your readers and enables you to view your paper differently. This can also help you gain some insight into any serious deficits that might require your attention.

c) As you read your first draft, look for logical flow of ideas and information from one section to another.

d) Ensure that areas that require particular attention and emphasis stand out, and compliment your paper.

e) These process (a-d) should be repeated several times until it is delightful to read the paper. The important message here is that if you do not like how it reads, others will not like it too. Note that at this stage, you can share the paper with colleagues for any constructive criticism or input.

f) Do the final checks to ensure that all aspects or requirements of the Journal or institution you are submitting to have been met.

References

1. Matthews, J.R. and Matthews, R.W. (2008) Successful Scientific Writing: a step-by-step guide for the biological and medical sciences. 3rd Ed. Cambridge University Press, Cambridge, UK.
2. Lai, J.C.K. (2002) Scientific Writing: A comprehensive introduction. *Unpublished guide.*

Chapter 14

THE FINAL CHECK

Afterthe rigorous task of compiling the paper and thorough editing, you need to take a break and cool off for a few days before resuming with the final check of all sections. This break is intended to allow the author reflect on what they have written, and can result in improved final revisions before submission. When doing the final check, pay particular attention to the following:

1. The title. Have a critical look at the title one more time and determine if it is clear and effective. This is the time to make any necessary final changes.
2. The abstract. Check for compliance to word limits, and missing or repeated information.
3. Check the organization of the figures and tables. Pay particular attention to the caption and legend.
4. Check and ensure that the most critical references have not been omitted.
5. Verify the overall compliance of the whole document with the desired Journal specifications. The January issue of the Journals publication usually contains instructions to authors. Go over them one more time just to ensure you have not missed anything. If the manuscript is in the wrong format, it will not be published.
6. Ensure that all pages have been numbered. Numbering will be useful to reviewers and editors in pointing out specific parts of your text for correction.
7. If there are no specifications, double space your entire document including legends for tables and figures, and footnotes. Margins must be wide (one inch at the sides and at minimum an inch at the top and bottom).

8. Some Journals require that the author indicates in the text of the document where the figures and tables will be placed. Check and ensure you have complied with that.

After the above checks, share the paper with the co-authors for one final round of review before you initiate the submission process. On occasion, serious errors can be spotted at this stage.

References

1. Matthews, J.R. and Matthews, R.W. (2008) Successful Scientific Writing: a step-by-step guide for the biological and medical sciences. 3rd Ed. Cambridge University Press, Cambridge, UK.

Chapter 15

THE SUBMISSION PROCESS FOR PUBLICATION

In my previous discussion in chapter one, I stated the importance of targeting your paper to a particular journal for consideration for publication. This is an important part of the planning process for writing a paper. It is quite common to find students and even scientists writing a complete paper without any target Journal in mind. This is a wrong approach, which later results in unnecessary waste of time in formatting and extensive editing.

With the tremendous internet resources available, most journals require online submission of manuscripts. Hence you will be required to register on the website of your target Journal to facilitate uploading of your paper and submission. It is not acceptable to send one paper to more than one journal at a time.

Once you have accomplished Journal targeting and requisite formatting, you need to write a cover letter to the Journal's editor that will be sent with your paper to the journal.

What does constitute the letter to the editor?

This letter should be very brief and must clearly state the compelling significance of your findings, which would be of interest to the respective journal and their readership. You must state whether it is a research paper, or a short communications. The letter to the editor serves another purpose; it gives the chief editor an idea of potential experts, who could be invited to review your manuscript. Note that it is acceptable to identify potential reviewers with conflict of interest who should not be invited to participate in the review of your paper.

Some journals may require a declaration of the contribution from each of the authors included in the cover letter.

If your paper is a follow-up to previous work that had been published with the target journal, state this fact in the letter to the editor. If the editor knows that you have previous work published with them, then you will have an advantage. This information is crucial even if the published work is in a different journal. Take note that this does not guarantee that your paper will be published, but does give you significant mileage.

Mistakes to avoid in the cover letter:

a) Do not forget to state the title of your paper and identify the corresponding author with full address and email/telephone.

b) Do not incorrectly spell the name of the chief editor or other editorial team member.

c) Do not forget to state the name of the journal you are submitting to.

d) Do not overly praise the Journal. However, it is acceptable to say nice things about that Journal in a moderate way.

e) A common mistake when resubmitting your article to a second journal after initial rejection is forgetting to remove the name of the previous Journal. Counter-check for this error.

With the paper and cover letter ready, you can start the submission process. It is important to take a lot of care, to ensure that you have uploaded all the required files before you hit the submission key. Whether submitting electronically or hard copy, you must always keep a back-up copy of the entire paper submitted. Strange things happen and things get lost at the most critical times.

Communication with the editorial board

It is always important to be patient with the editorial team when they take time to give feedback on your submission. However, if you have not heard from the editorial board after six weeks or so, send an inquiry through an email or phone call. I have noted, for some Journals, you get faster feedback to such inquiry if you send it through the Journal manuscript submission portal while signed in. In all your communications with the editorial board, have great etiquette and maintain a professional and patient disposition. Most of the time, they have a lot in their hands.

What next if the paper needs major corrections or is rejected?

The best scenario for any author is when the chief editor communicates indicating that the paper has been accepted for publication or requires minor revisions to address the reviewer's concerns. However, this may not always be the case. In competitive, and reputable scientific journals, it is quite normal to have a paper returned for major corrections or out-rightly rejected. This is not an easy event for the authors. Without any doubt, this can be painful, especially if it is the first attempt to publish. My advice for you is to take in your stride and learn from it. If you truly dislike the reviewers' comments for whatever reason, wait for at least two days before any response. This will allow you time to cool off, and avoid unnecessary and emotional tirade. There is a saying that you do not discipline a child when angry. However, never ignore the detailed response from the reviewers and the editor. Use these comments to improve the paper and increase chances for publication in the subsequent attempt.

When a paper is rejected, the author is allowed to respectively appeal the decision to the editorial board. Alternatively, you can re-submit to the same Journal if the chief editor is willing to reconsider the paper. Otherwise, re-format it to fit specifications for another Journal and submit afresh.

Your second re-submission to the same Journal requires a rebuttal or point by point response to the queries raised by the reviewers. You must respond to all the queries, and even those you disagree with, giving justifications.

Note that it is in order to disagree with some reviewers' comments. While reviewers' comments or concerns should be used to improve your paper, they might not be spot on all the time. Learn to carefully evaluate such criticism, and sieve out what you do not need or disagree with it. It is always useful to discuss the reviewer's comments with your colleagues or supervisors in your scheduled laboratory meetings. Your colleagues might, for example help you appreciate a positive aspect of the reviewer's comments, initially perceived negatively.

I would like to unequivocally state that the rejection of a paper does not always imply poor science. You will be interested to know that manuscripts with pioneering work and remarkable breakthroughs were rejected by publishers. Sometimes you have to trust your gut feelings, especially if your research is reporting findings radically different from the popular understanding. Of course the risk is there and a bold approach to science can be career-ending when proven wrong. But again, history has taught us that courage in science, spurs new discoveries with potential for great social and economic change.

References

1. Dizon, A.E. and Rosenberg, J.E. (1990) We Don't Care, Professor Einstein, the Instructions to the Authors specifically said double spaces. In writing for Fishery Journals, Ed.J. Hunter. Bethseda, MD: American Fisheries Society.

2. Matthews, J.R. and Matthews, R.W. (2008) Successful Scientific Writing: a step-by-step guide for the biological and medical sciences. 3rd Ed. Cambridge University Press, Cambridge, UK.

Chapter 16

ETHICS IN RESEARCH

E thics in research is a wide subject covered by many books. However, in scientific writing, the biggest culprits are usually plagiarism and falsification of data. Scientific discovery, advancement and excellence depends a lot on honesty and integrity of scientists. This is an expectation that must be met by all scientists at all costs without any exception whatsoever. Ensure that, at all times, you give credit to others fully when making references to their work. In science, deliberate plagiarism is the master of all malpractice and is immoral and intellectual dirt. The issue of data manipulation and plagiarism is taken so seriously in science because accuracy and honesty forms the foundation on which science is anchored.

One other thing to avoid is self-plagiarism. This refers to recycling of data that is adequate for one paper, to produce two or more, with a few additions of new information here and there. While this is not illegal, it undermines science and is a waste of time for many people involved in the review process.

In situations where a scientist is reporting research conducted in animals or human beings, evidence of having been granted institutional ethical approval is required. This must be provided as a scanned copy or a clear statement indicating that such approval was granted.

While this book has not discussed ethics in research in great detail, it has mentioned in brief what is important in regard to the writing process.

References

1. Buranen, L. and Roy, A.M. (eds.) (1999) Perspectives on Plagiarism and Intellectual Property in a postmodern World. Albany: State University of New York Press.
2. Matthews, J.R. and Matthews, R.W. (2008) Successful Scientific Writing: a step-by-step guide for the biological and medical sciences. 3rd Ed. Cambridge University Press, Cambridge, UK.

SELECTED REFERENCE LIST

1. Alley, M. (1996) The craft of scientific writing, 3rd Ed. Prentice Hall NJ.
2. Bem, D.J. (1987) Writing the empirical Journal Article. In M.P. Zanna and J.M. Darley (Eds.). The complete academic: A practical guide for the beginning social scientist (pp. 171-201) New York: Random House.
3. Benson, B.W. and Boege, S. (2002) Handbook of Good Laboratory Practices. Bristol, PA: Hemisphere Publishing.
4. Buranen, L. and Roy, A.M. (eds.) (1999) Perspectives on Plagiarism and Intellectual Property in a postmodern World. Albany: State University of New York Press.
5. Council of Biology Editors Style Manual Committee (1994) Scientific Style and Format: The CBE Manual for Authors, Editors, and Publishers, 6th Ed., Cambridge and New York.
6. Council of Science Editors (2006) Scientific Style and Format: The CSE Style Manual for Authors, Editors and Publishers. 8th Ed. New York: Cambridge University Press.
7. Creswell, J.W. (2009) Research Design: qualitative, quantitative and mixed methods approaches. Third Edition. SAGE publications Ltd, Thousand Oaks California.
8. Day, R.A. (1998) How to write and publish a scientific paper. 5th Ed. Orynx press.
9. Day, R.A. (1994) How to write and publish a scientific paper. Cambridge University Press.
10. Dizon, A.E. and Rosenberg, J.E. (1990) We Don't Care, Professor Einstein, the Instructions to the Authors specifically said double spaces. In writing for Fishery Journals, Ed.J. Hunter. Bethseda, MD: American Fisheries Society.

11. Dodd, J.S. (1986) The ACS Style Guide: A Manual for Authors and Editors, American Chemical Society, Washington DC.

12. Goben, G., and Swan, J. (1990) The science of scientific writing. Am. Scientist 78:550-558.

13. Hailman, J.P. and Strier, K.P. (1997) Planning, Proposing, and Presenting Science Effectively. Cambridge UK: Cambridge University Press.

14. Isaac, A.O., Dukhande, V.V., and Lai, J.C. (2007) Metabolic and antioxidant system alterations in an astrocytoma cell line challenged with mitochondrial DNA deletion. Neurochem Res. 32(11):1906-18

15. Isaac, A.O., Kawikova, I.B., Daniels, C.K., Lai, J.C. (2006) Manganese treatment modulates the expression of peroxisome proliferator-activated receptors in astrocytoma and neuroblastoma cells. Neurochem Res. 31(11):1305-16.

16. Iverson, C.I. (1997) American Medical Association Manual of style: A Guide for Authors and Editors. 10th Ed. Baltimore MD: Lippincott, Williams and Wilkins.

17. Kothari, C.R. (2006) Research Methodology: Methods and techniques. *New Age International (P) Limited, Publishers.*

18. Lai, J.C.K. (2002) Scientific Writing: A comprehensive introduction *(Unpublished Guide).*

19. Mathews, J.R., Bowen, J.M., and Mathews RW (1996) Successful Scientific Writing: A step by step Guide for Biomedical Scientists, Cambridge University, Cambridge U.K.

20. Matthews, J.R. and Matthews, R.W. (2008) Successful Scientific Writing: a step-by-step guide for the biological and medical sciences. 3rd Ed. Cambridge University Press, Cambridge, UK.

21. Maxwell, J.A. (2005) Qualitative research design: An interactive Approach (2nd Ed). Thousand Oaks, CA: SAGE.

22. McMillan, V. (1988) Writing papers in the biological sciences. Bedford books. NY.
23. Mugenda, O.M. and Mugenda. A.G. (2003) Research Methods: Quantitative and qualitative approaches. Acts Press.
24. Ogechi, S.N. (2013) Research Methods: Question and Answer Revision Book.
25. Publication Manual of the American psychological association (2001) 6th Ed. Washington DC: American Psychological Association.
26. Rubens, P. (2002) Science and Technical writing. A manual of style. 2nd Ed. Oxford: *Routledge study guides.*
27. The Basics of Scientific Writing in APA style: Manual of The American Psychological Association (6th Ed.) 2010.
28. The International Committee of Medical Journal Editors (2006)
29. Wilkinson, A.M. (1991) The scientists handbook for writing papers and dissertations, Prentice Hall, Englewood Cliffs, NJ.
30. Wilkinson, A.M. (1991) The Scientists handbook for writing papers, dissertations. Eaglewood Cliffs, NJ: Prentice Hall.
31. Wilkinson, A.M. (1991) The scientists Handbook for writing papers and dissertations, Prentice Hall, Englewood Cliffs, NJ.
32. Zeiger, M. (1991) Writing biomedical research papers, McGraw-Hill, New York.
33. Lebrun, J.L. (2007) Scientific Writing: A Reader and Writers Guide. World Scientific Publishing Co Pte Ltd, Singapore.

INDEX

APPENDIX 1
Some standard abbreviations

Adenine, Thymidine, Guanine, Cytosine, Uracil - A,T,G,C,U
Base pairs - bp
Carbon, nitrogen, phosphorus - C,N,P
Deoxyribonucleic acid - DNA
Deoxyribonucleotide triphosphate - dNTP
Ethylene glycol-bis(b-aminoether) N,N,N',N'-tetra acetic acid - EGTA
Ethylenediamine tetraacetic acid - EDTA
Grams - g
Hours - hr
Kilobase pairs - Kb
Litre - L
Mega base pairs - Mb
Micrograms - μg
Microliter - μl
Milligrams - mg
Milliliter - ml
Minutes - min
Nicotinamine adenine dinucleotide - NAD
Nucleotide triphosphate - NTP
Plaque forming units - PFU
Ribonucleic acid - RNA
Seconds - sec
Tris(hydroxyamino) methane - TRIS
Ultraviolet light - UV

Amino acids can be abbreviated using one or three letter abbreviations

Ala (A)	Alanine
Arg (R)	Arginine
Asn (N)	Asparagine
Asp (D)	Aspartic acid
Cys (C)	Cysteine
Gln (Q)	Glutamine
Glu (E)	Glutamic acid
Gly (G)	Glycine
His (H)	Histidine
Ile (I0)	Isoleucine
Leu (L0)	Leucine
Lys (K)	Lysine
Met (M)	Methionine
Phe (F)	Phenylalanine
Pro (P)	Proline
Ser (S)	Serine
Thr (T)	Threonine
Trp (W)	Tryptophan
Tyr (Y)	Tyrosine
Val (V)	Valine

ABOUT THE AUTHOR

Alfred Orina Isaac, PhD

Dr. Isaac attended Idaho State University for his doctoral studies (Pharmaceutical Sciences) and then proceeded to Case Western Reserve University Medical Center for post-doctoral studies in neuroscience. Currently, Dr. Isaac is the Director, School of Health Sciences and Technology, at The Technical University of Kenya. Dr. Isaac is a Pharmaceutical Scientist with specific interest in neuroscience. To this end, Dr. Isaac has published and contributed in the field of neurotoxicology and neuroprotection. In addition to research and publications, Dr. Isaac has mentored and supervised many undergraduate and post-graduate students.

CPSIA information can be obtained
at www.ICGtesting.com
Printed in the USA
BVHW040219220819
556515BV00014B/361/P

9 781514 289617